AI, Bitcoin, & Nostr for Women

Also from EATMS Productions

Books on power, survival, women's autonomy, and the systems shaping modern America.

Nonfiction

Billionaires, Capitalism, and Power

Evil and the Mountain Ungreed
Self Help for American Billionaires
Selfish Steve and the Ivory Tower
Tariffs, Taxes, & Face-Eating Leopards
Ban Billionaires: Fascism Fix

Fascism, Religion, and Cultural Control

Self Help for the Manosphere
Fascism 2025
Fascism & the Perverts & the Greed Virus
Christian Fascism Marriage Book
Tyranny, Table Manners, & Tiramisu

Guides for Women's Autonomy and Protection

How to Survive in Post-America as a Woman
Project 2025 American Drag
4B – Burn, Ban, Boycott, Build
4B OG – So No Go GYN
I'm Glad He's Dead

Analysis of Authoritarian Project 2025

Project 2025: The Blueprint
Project 2025: The List
Project 2025, Christian Dumb Dumbs, & The Republican Agenda
Fascism, Project 2025, & The Pinkprint

Modern Rewrites for Women

Stoic Principles Reimagined
Siddhartha Reimagined
The Prince Reimagined for Women
The Art of War Reimagined for Women
The Jungle Reimagined
The Constitution Reimagined for Women

Machine Learning Series

AI, Bitcoin, Nostr for Women
AI, Safety, & Security for Women
AI, Anxiety, & Health for Women
AI, Kids, & Family Safety for Women
AI, Creativity, & Personal Expression for Women
AI, Independent Work, & Parallel Power for Women

Social Systems Series

Emotional Labor for Women
Household Power for Women
Workplace Power for Women
Medical Bias for Women
Aging Systems for Women
Recovery Systems for Women

Fiction

Dystopian Stories of Resistance and Collapse

Propaganda Paige & the Missing Prosperity
Propaganda Paige & the TIDE Manifesto
Propaganda Paige & the Shadow Cartographers
Propaganda Paige & the Prosperity Alliance
Propaganda Paige & the Shattered Truth
Propaganda Paige & the Rising TIDE
Propaganda Paige & the Last Bastion
Propaganda Paige & the Dawn of Prosperity
Project 2025: Dorian — The Last Men
Project 2025: Boy — A Last Men Novel

AI, Bitcoin, & Nostr for Women

A Survival Guide For Women in Authoritarian America

by
Mads Duchamp

Foreword
by Esme Mees

EATMS
PRODUCTIONS

Copyright © 2025 Eatms Productions
All rights reserved.

No part of this book may be reproduced, or stored in a retrieval system, or transmitted in any form or by any means, electronic, mechanical, photocopying, recording, or otherwise, without express permission in writing from the publisher.

This book is a work of opinion and creative interpretation. While some names and events may be referenced or alluded to, any claims made are based on publicly available information and are intended as satire, parody, or commentary on societal and political issues. The content should not be interpreted as factual assertions about any individual or entity. The author does not intend to defraud, defame, or mislead, and encourages readers to form their own conclusions. Any resemblance to real persons, living or dead, is purely coincidental unless explicitly noted otherwise.

ISBN: 978-1-966014-33-1

Cover, interior design by: Esme Mees

  eatms@pm.me
  www.eatms.me

Check out EATMS Underground:
https://tinyurl.com/eatmsNOSTR

Printed in the United States of America.

I hope to bequeath to the future a calculus of the nervous system.

— Ada Lovelace

Table of Contents

Foreword	11
Chapter 1—The New Power Landscape	13
Chapter 2—AI: What It Actually Means for You	25
Chapter 3—Bitcoin Without the Bros	35
Chapter 4—Nostr: Speech, Identity, & Escape Routes	47
Chapter 5—Digital Self-Defense for Women	57
Chapter 6—Building Parallel Power	69
Chapter 7—Crisis Mode: When Systems Start Closing	78
Chapter 8—The EATMS 7-Day Starter Plan	90
Conclusion	98
About the author	102

Foreword

Women are told they are free, yet the systems around them grow more controlling every year. Technology promises connection and safety, but it often works as a form of quiet surveillance. Government agencies, corporate platforms, and automated systems collect our data, judge our choices, and make decisions about us without ever seeing our real lives. These structures were not built with women's needs in mind. They were built for efficiency, compliance, and profit. When women move through them, they are often misread, dismissed, or pushed aside.

This book is meant to shift that balance. It explains how the modern digital world actually operates and gives you practical steps for protecting your privacy, your money, and your voice. You do not need technical skill to stay safe. You only need clear explanations and honest guidance, tools that help you see the patterns of control instead of being trapped in them. Once you understand how these systems function, you can navigate them with confidence instead of confusion.

I wrote this foreword because women deserve more than fear or vague advice. They deserve strategy and straightforward tools that respect their intelligence. This book offers exactly that. My hope is that by the time you finish, you feel more prepared, more steady, and far less willing to hand your autonomy to any institution, corporate or governmental, that was never designed to protect it.

—Esme Mees, Winter 2025

~1
The New Power Landscape

Over the past twenty years, life in the United States has been pulled into digital systems that most people never chose. Banking, healthcare, communication, shopping, work, and even personal identity now pass through a handful of powerful companies and the software they control. This shift did not happen through public debate or voting. It happened quietly, through new apps, required logins, automated services, and the steady push toward "smart" everything. People accepted these tools because they were convenient. Few noticed that, with each step, more authority moved into private systems that answer to profit and control rather than the needs of ordinary women.

Artificial intelligence has become the hidden engine behind many decisions that shape daily life. These systems analyze huge amounts of data and make choices about credit, jobs, insurance, school records, and even safety risks. Financial companies have grown larger and fewer, which means they now have more power over who gets access to money and who does not. Social media platforms have replaced newspapers and town meetings as the main place where people speak publicly, but these platforms are owned by corporations that decide what posts get seen and what gets buried. Meanwhile, every search, purchase, and message creates data that can be tracked, stored, and sold. Most people do not see any of this directly. They only feel the impact when something in their life is suddenly denied, flagged, or restricted with no explanation.

Women feel these changes more strongly than men because they interact with more systems in daily life. They schedule medical visits for family members, manage school accounts, balance work and home, search for services, and communicate across many platforms. All of these tasks now pass through digital systems that collect information. These systems were not designed with women's realities in mind. They often treat caregiving gaps, part-time work, job changes, or increased communication as signs of weakness or instability. The system reads women's data as if they were problems to solve rather than people with complex lives.

Speech platforms also shape women's experiences in ways that are hard to see at first. A woman who posts about safety, harassment, unfair treatment, or personal struggle may find her content hidden or removed. This happens because automated moderation tools look for keywords, not context. The system cannot tell the difference between a person describing harm and a person causing harm. This pushes many women to stay quiet, not because they want to, but because speaking honestly can lead to penalties. Over time, public conversation becomes shaped by what the system allows, not by what women need to say.

Financial systems add another layer. Women often have work histories that include breaks, part-time shifts, or changes related to caregiving. Automated scoring tools do not understand this. They only see gaps. They treat these gaps as signs that a woman is less reliable, even when she is carrying more responsibilities than the model can measure. This can lead to lower credit scores, denied loans, or reduced access to resources. The system is not judging her character. It is judging patterns it never learned to read correctly.

All of these forces create a kind of pressure that is easy to overlook because it does not look dramatic. There are no public warnings, no loud threats, and no obvious signs of control. The systems are smooth and polite. They ask users to "agree to terms," offer quick choices, and present automated decisions as facts. Yet each step pulls more power away from individuals and gives more authority to institutions that are not accountable to the people they affect.

The first section of this chapter shows how these systems rose around us and why they affect women in particular. The next section will explain how these systems sort, label, and control outcomes, often without the user's knowledge. It will break down how profiling already happens, why it matters, and what women need to understand before they can push back.

The systems that shape daily life feel neutral on the surface, but they are not neutral in practice. Artificial intelligence, banking tools, speech platforms, and data trackers all make decisions based on patterns taken from the past. Those patterns carry the same inequalities that already existed. When these systems run on autopilot, they do not fix old problems. They repeat them. For women, this means the challenges they face in the real world become multiplied inside the digital world, often without warning.

Artificial intelligence is a good example. Companies use AI to sort job applications, judge medical needs, check insurance risks, and even predict who might struggle with debt. These systems rely on old data, and old data reflects decades of bias. Women with caregiving breaks, part-time schedules, or nontraditional work histories often confuse the models. The system reads these patterns as signs of instability, even when the woman is responsible, highly skilled, and managing more duties than the model can understand. A gap in work may reflect raising a child or caring for a parent, but AI treats the gap as a flaw. Because the system cannot read context, it often labels women as less dependable even when the opposite is true.

The same problem shows up in financial systems. Banks and credit companies use automated scoring tools that reward steady work, predictable income, and uninterrupted employment. Many women do not fit that pattern because their lives include caregiving demands, job changes, health needs, or shifts in responsibility. The models do not understand why these changes happen. They only react to the fact that the changes exist. As a result, a woman may be denied a loan, charged higher interest, or given a lower credit score even when she manages her finances well. This is not because someone dislikes her. It is because the system reads her life as an exception, and exceptions are treated as risks.

Speech platforms create another layer of pressure. Women who talk openly about difficult issues often trigger automated filters that cannot tell the difference between reporting harm and causing harm. A message about domestic violence, workplace abuse, political events, or fear can be flagged as unsafe content. Posts may be hidden, removed, or buried so deep that almost no one sees them. This does not happen because the platform wants to silence women. It happens because the system is built to avoid anything that might scare advertisers or create controversy. Women learn to stay quiet not because they lack opinions, but because telling the truth often leads to invisible consequences.

Surveillance makes the situation more complicated. Every purchase, every message, every appointment, and every search produces data. Women generate more of this data because they often manage the needs of children, aging parents, households, and workplaces. The system interprets this flood of information without knowing anything about the emotional, physical, or financial labor behind it. A woman dealing with medical visits or school messages may appear stressed, unstable, or disorganized in the data. The system does not see the care and responsibility involved. It only sees patterns and assigns meaning to them.

All of these systems combine to create a kind of quiet pressure that shapes a woman's choices long before she notices anything is wrong. A job application disappears without explanation. A post she cares about never gains visibility. A bank account is flagged for review. A service denies her a benefit. Each moment feels small, but together they create a pattern that limits her independence. The pressure is real even if it is hidden.

These forces matter because they influence opportunities, finances, safety, and identity. They affect who is heard, who is trusted, who gets access, and who gets ignored. Women often discover these limits only after a decision has already been made, long after the system has labeled them in some way. Understanding these mechanisms is the first step toward regaining control.

The digital systems that now shape daily life do not announce their power or explain their decisions. They work in the background, making choices that affect money, opportunity, safety, and public voice. Many people assume these systems are fair because they feel modern and efficient, but the truth is simpler. These systems make fast decisions, not thoughtful ones. They copy patterns from the past instead of understanding the present. They reward people who fit narrow expectations and push aside anyone who does not. Women feel the effects first because their responsibilities and routines do not match the patterns the system prefers.

When a woman is judged by an algorithm, the decision often becomes permanent. A lower credit score, a flagged account, or a rejected job application can follow her into future systems because her data travels farther than she does. When a post she writes gets removed or buried, the platform does not explain what happened or how to avoid it next time. When her digital footprint grows from managing her family, her work, and her home, that data can be combined in ways that give a false picture of her life. These systems are not designed to see the full story. They only see fragments, and from those fragments they make decisions that can shape her future.

The most important thing to understand is that these systems are already active. They are not predictions about what could happen one day. They are shaping outcomes now, often without the user's knowledge. A woman may feel like she is making free choices, but she is often responding to limits she did not create. The system may label her as unreliable when she is actually responsible. It may treat her as high risk when she is simply carrying more duties than the model can measure. It may silence her when she tries to speak honestly. The system's mistakes feel personal even when they come from cold calculations.

Recognizing this pressure is not meant to create fear. It is meant to create clarity. A woman cannot protect her independence if she cannot see where her power is being drained. She cannot make informed decisions if she believes the system is fair when it is not. She cannot navigate the digital world safely if she thinks errors are her fault rather than the result of faulty design. Understanding these systems is the first step toward pushing back. The next chapters show how to build that resistance with real tools rather than empty talk.

The summary below captures the main points where women lose power in the default system and why these points matter. It is not a full map, but it is enough to show where attention needs to shift if women want to keep or regain control.

Women lose power first at the moment they are classified. Algorithms place women into categories based on limited data, and those categories shape future decisions. A caregiving gap or a change in schedule may be read as instability, even when the woman is doing the work that holds others together. Women also lose power in public expression. When automated systems hide or remove posts about danger, harm, or inequality, women lose visibility and learn to silence themselves. Financial systems create another point of loss by judging women according to models built on male patterns of work.

These systems do not see the unpaid labor women perform, and they punish any break in a traditional career path. Women also lose power when their digital footprint grows without context. The system collects large amounts of data from their daily responsibilities but does not understand what the data means. This leads to false assumptions and unfair labels. The final point of power loss is the disappearance of appeal. Automated systems rarely allow users to challenge or correct a decision. When a woman is mislabeled or denied something important, she is often left with no one to talk to and no way to fix the mistake.

There are several clear warning signs that show when these systems are starting to shape a woman's life. One warning sign appears when she begins to believe that automated decisions are always fair or correct. Another sign appears when she notices her online posts reaching fewer people, which may mean the platform's filters are working against her. A third warning sign is a sudden financial block, such as a rejected application or frozen account, which often comes from automated scoring rather than human review. A fourth warning sign shows up when she realizes her digital footprint has grown from managing the needs of others, making her easy to track or misjudge. The final warning sign is the belief that avoiding technology will protect her. Opting out does not stop the system from gathering data about her. It only means she has less control over what the system learns.

This chapter maps the landscape of control. The chapters that follow show how to move through it with intention instead of fear. They explain how to use artificial intelligence without feeding systems that already misread women, how Bitcoin can create financial safety outside banks that track and judge, and how Nostr offers a place to speak without corporate or governmental filters. The goal is not to run from the modern world. It is to understand its pressures and to build enough independence that no system, public or private, gets to decide your limits.

Summary Chart: Where Women Lost Power in the Default System

Classification
Algorithms often misread caregiving breaks, part-time work, or changing schedules as signs of weakness. Without context, the system labels women as unreliable, leading to fewer interviews and lower chances for advancement.

Visibility
Automatic filters bury posts about safety, harm, or inequality because they look risky to advertisers. Women's voices are pushed out of view, and many learn to stay quiet to avoid penalties.

Financial Evaluation
Scoring tools judge women by models built around steady work patterns that many do not have. This leads to unfair credit scores, denied loans, or higher rates, even when a woman manages her finances well.

Data Exposure
Daily tasks like medical visits, school messages, and household planning create large data trails. Companies use this data to predict behavior, often in ways that misunderstand a woman's real situation.

Loss of Appeal
When automated systems make a bad decision, it is often final. There is no one to explain the issue to and no way to correct the system. These errors can follow a woman into future decisions and limit her options.

5 Quick Risk Indicators

1. Believing automated decisions are always fair.
   When a woman sees an algorithm's judgment as neutral, she may accept an outcome that is biased or based on incomplete information.
2. Speaking openly online without knowing how moderation works. Posts about safety, harm, or inequality are often hidden or removed by automatic filters, even when the content is truthful.
3. Depending fully on traditional financial systems.
   A single automated check can freeze an account or block access to money, and there is often no person available to fix the mistake quickly.
4. Letting a large digital trail build up without noticing. Every app, form, and account creates data that companies can analyze. These patterns can be misread and used to judge her unfairly.
5. Thinking avoidance equals protection.
   Staying off platforms does not stop the system from collecting information. It only means she has less control over how she is seen or judged.

Bonus Tip: Remember Who Built These Systems, and Why

Most digital systems were designed by men, built by male-led companies, and shaped by goals that rarely include women's needs. These systems collect data, sort people, and push behavior because that is how they make money and maintain control. But understanding this does not mean giving up or stepping back. It means entering these spaces with your eyes open. With a little education, patience, and steady practice, women can learn how these tools work and use them for their own benefit. The more we understand the system, the less power it has over us. The goal is not to fear technology, but to use it in a way that protects our freedom and strengthens our place in the world.

~2
AI: What It Actually Means for You

Artificial intelligence has moved from the edges of digital life to the center of everyday decision-making. Many people still imagine AI as something used only by researchers or large companies, but it now influences choices that affect almost everyone. When you apply for a job, seek medical care, request help from a school, ask for a loan, or post online, an AI system is often the first to judge your information. These systems do their work quietly and without explanation. Most people have no idea how often a machine, not a human, decides what they deserve, what they can access, or how they will be classified.

AI is designed to make decisions quickly by finding patterns in large amounts of data. The problem is that most of the data it uses comes from the past. That means AI often copies old biases and mistakes. If women were treated unfairly in older hiring systems, the AI trained on that history will repeat the unfairness. If women's medical symptoms were dismissed or misunderstood for decades, the AI that learns from old medical records will make the same errors. If certain communities were policed more heavily in the past, AI used in modern policing may continue those patterns even when nothing in the present justifies it.

AI also affects people in quieter ways. Many customer service systems no longer give you a person to speak to because an automated tool has already made a decision about your case. Schools use AI to flag "concerning" behavior in student messages. Hospitals use AI to decide who gets fast attention and who waits. Banks use AI to decide whether your spending pattern looks normal or suspicious. Even dating apps and social platforms use AI to decide who sees you and who does not.

For women, the impact is deeper because their lives rarely follow the simple patterns these systems expect. Caregiving gaps, flexible schedules, emotional labor, and multitasking do not fit easily into AI's models. A woman may look inconsistent to a system that expects steady work hours or a predictable routine. She may look unstable to a system that does not understand the stress she carries. She may look less serious to a system that misreads tone or context. AI does not mean to get it wrong. It simply cannot read the full truth.

The influence of AI grows each year, and it is unlikely to slow down. This does not mean women should fear it or avoid it. AI can be a powerful tool when used with awareness. It can help with planning, learning, financial decisions, and creative work. It can help women find information faster or build projects they would not have had time to do alone. The key is understanding where AI helps and where it harms, and how to use the tool without letting it define you.

This chapter describes the terrain. The chapters that follow explain how to move through it. They show how artificial intelligence can be used without feeding it private information, how Bitcoin can create financial safety outside traditional banks, and how Nostr can provide a space to speak without being policed by corporate platforms. The goal is not to escape the modern world but to learn how to live in it with more control and less fear.

The next section will explain how AI is used in specific areas like hiring, healthcare, social services, policing, and credit systems. It will show how these models make decisions, why they often misunderstand women, and what signs to look for when an AI system is shaping something important in your life.

AI plays a major role in hiring, yet most applicants never know it. Many companies use computer systems to scan resumes and applications before a human ever sees them. These systems look for certain keywords, timelines, and patterns. A straight career path with no breaks is often rewarded because it fits what the model learned from past successful applicants. A woman with caregiving gaps or a part-time job mixed into her work history can be pushed aside even if she is fully qualified. The AI does not know why a break exists. It only measures the break itself. As a result, many women are filtered out long before they have a chance to explain their skills or experience.

AI also appears in healthcare, where it helps hospitals decide who gets fast attention and who can wait. These systems learn from years of medical records, but those records contain bias. Women's pain has been dismissed, underdiagnosed, or labeled as emotional for generations. Because of this, an AI trained on old data often repeats the same mistakes. Women can be ranked lower in urgency or taken less seriously even when their symptoms are real and dangerous. AI does not make this choice on purpose, but the patterns it follows are shaped by history, not fairness.

In social services and education, AI tools are used to watch for "risk." Schools use software to scan student messages and search terms. Welfare programs use predictive tools to flag applicants for extra review. These systems often misread stress, poverty, or complex home lives as danger. Women who manage children, aging parents, or unstable living situations can be targeted by automated suspicion even when they are doing everything right. AI turns daily struggle into a red flag because it does not understand context.

Policing has also changed because of AI. Many departments use predictive software to decide where to patrol or whom to watch more closely. These models learn from old arrest data, and if certain neighborhoods were over-policed in the past, the AI continues that pattern. Women living in these areas may find themselves under more scrutiny for simple daily tasks. Something as small as walking with a phone, sitting in a parked car, or checking on a neighbor can be labeled suspicious. The system is not reading behavior. It is reading patterns shaped by past injustice.

Credit and financial systems rely heavily on AI as well. Banks use algorithms to judge spending habits and decide whether someone is a risk. A woman with shifting work hours, irregular paychecks, or caregiving expenses may look less stable to the system even when she manages her money carefully. A single automated flag can freeze an account until a review is done. These reviews often move slowly because there is no person watching closely. The AI makes the judgment, and the user must wait.

All of these examples show one clear pattern. AI does not understand the reality of women's lives. It only understands the data it has seen before. When women are treated unfairly by AI, it is not because the system is malicious. It is because the system is blind. It learns from a past that did not treat women equally, and then it repeats that past in new ways.

The next section will explain how women can use AI safely, how to reduce the amount of personal information the systems collect, and how to make AI work for them rather than against them. It will also include the ten AI privacy habits, safer-use scripts, and an example of a simple AI audit of a woman's digital life.

AI is a powerful tool, and it can help women save time, learn new skills, manage information, and build independence. But like any tool, it must be used with awareness. When a woman uses AI without thinking about privacy or data safety, she can give away more than she intends. When she uses it with clear limits, she can get help without losing control of her personal life. The goal is not to avoid AI or fear it. The goal is to use it in a way that protects her identity, her choices, and her long-term security.

Most AI systems collect information by studying what people type, upload, or search. Some systems store data for long periods, and others use it to train new models.

A woman does not need to understand the technical details to protect herself. She only needs to understand the basics. Anything shared directly with AI may be stored, connected to other data, or used to form a profile of her interests, routines, and concerns. The less personal detail she gives, the safer she becomes. AI can still be useful without knowing everything about her.

Using AI wisely means keeping firm boundaries. A woman can let AI help with research, planning, writing, or learning, but she should not give it anything that exposes her private life. That means no medical details, no financial accounts, no family information, and no personal history she would not want stored or analyzed. Creating a separate email or login for AI tasks can add another layer of safety. Some women even use a separate device for work that involves sensitive topics. These small choices let her benefit from the technology without handing over parts of her life to systems that do not answer to her.

10 AI Privacy Habits

1. Use a separate email address for AI tools so they are not connected to your main accounts.
2. Avoid sharing medical, financial, or family details unless the tool clearly states it does not store them.
3. Delete chat histories regularly when the platform allows it.
4. Turn off data-saving or training options in AI app settings whenever possible.
5. Use AI for tasks, not for confessions. Keep personal emotional information offline.
6. Read the privacy page before using a new AI tool, even if you only check the key points.
7. Never upload IDs, documents, or photos that reveal your exact identity.
8. Use a VPN when doing sensitive searches to reduce tracking.
9. Keep your device updated so security holes are patched.
10. Treat AI as public space. If you would not post it online, do not put it into a model.

Scripts for Safer AI Use

Script for asking sensitive questions safely:
"I need general advice on a situation. Please answer without needing personal details."

Script for blocking data requests:
"I prefer not to share identifying information. Give me guidance based on general examples."

Script for refusing to reveal family or medical details:
"I cannot provide specifics. Give me the steps someone in this situation should follow."

Script for financial questions:
"I want general financial guidance without linking this to my accounts or personal data."

Script for professional or hiring questions:
"Explain best practices for someone navigating this situation without using personal history."

These scripts create distance and reduce the amount of information AI can store about your life.

Example: A Simple AI Audit of Your Digital Life

A woman who wants to understand her AI exposure can do a quick review of her digital routine. She begins by listing the apps and websites she uses each day. She checks which of them use AI for sorting, recommendations, ads, customer service, or decision-making. She notes how often she types personal information into these systems. She reviews whether her phone stores voice recordings, whether her email scans messages for keywords, and whether her search engine tracks location. She looks at her cloud storage to see if private files are backed up in places she did not expect.

Next, she reviews where she has used AI directly. She checks old chat logs, uploads, and shared documents. She scans for anything that contains names, dates, addresses, medical information, or financial records. If she finds anything sensitive, she deletes it and switches to safer habits. She then adjusts privacy settings, turns off tracking features, and begins using AI with clearer boundaries. This quick audit gives her a realistic picture of what AI knows about her and what she wants to keep private going forward.

Bonus Tip: Protect Your Health and Sexuality Data

Many apps quietly track period dates, sexual activity, symptoms, pregnancy status, and location. Some of this data can be shared or sold to companies that use AI to predict behavior. To stay safer, avoid putting private health information into apps you do not fully trust. Turn off location services for health apps, and delete any app that asks for details it does not need. If you have questions about your body, use a private browser or a trusted adult instead of platforms that collect data. The less personal information you feed into these systems, the less they can store, link, or use against you later.

~3
Bitcoin Without The Bros

Bitcoin is one of the most talked-about technologies in the world, yet still one of the most confusing. Many people hear about it through hype or fear, and others hear about it from men who treat it like a competition instead of a tool. In reality, Bitcoin is simply another kind of money. It runs digitally, it moves without banks, and it can be held directly by the person who owns it. You do not need technical skill to understand it. You only need to see what problems it was built to solve and why it matters in a world where financial systems are becoming more restrictive and less predictable.

Bitcoin matters because it allows people to store and move value without relying on banks or payment companies. Many banks freeze accounts, delay transfers, or demand documents at the worst possible moments. Bitcoin gives the user direct control as long as she protects her keys and follows basic safety steps. This makes it useful in moments of financial stress, unstable work schedules, or emergencies. It can also help women who shift between jobs, freelance, or manage households where money comes from multiple sources. In these situations, having a form of money that does not depend on a single institution can create real stability.

Women can gain a lot from Bitcoin, but many are pushed away by the culture that surrounds it. Online spaces focused on Bitcoin often reward aggressive behavior, risky thinking, and a loud style of debate. These environments can feel hostile or overwhelming, especially for women who are looking for practical guidance rather than ideology. The loudest voices talk about getting rich or beating the system, but those goals ignore the real value Bitcoin offers. It is not a path to instant wealth. It is a way to hold money that is harder to seize, freeze, or quietly monitor.

It is also important to understand that Bitcoin is not perfect. It does not fix every financial problem and does not replace the need for smart planning. It is not fully anonymous, and people who use it carelessly can expose more information than they expect. But when used carefully, Bitcoin can offer more privacy and independence than regular banking tools. It can help women keep a small emergency fund that no one else controls, build savings outside shared accounts, or protect money in times of political or economic instability. It can also help women who face financial abuse or want a safe way to create distance and prepare for change.

The key idea is simple. Bitcoin is neither a miracle nor a danger by itself. It is a tool, and tools become powerful when the user understands the basics and sets clear boundaries.

The key idea is simple. Bitcoin is neither a miracle nor a danger by itself. It is a tool, and tools become powerful when the user understands the basics and sets clear boundaries. The next section will show how Bitcoin works in everyday life, how to choose a safe wallet, how to move money without mistakes, and how to avoid the traps that cause most new users to lose control.

Bitcoin becomes easier to understand once you see how it works in everyday life. At the center of Bitcoin is something called a wallet. A wallet is not a bank account and it is not stored at a bank. It is simply a small piece of software that holds the information you need to send and receive Bitcoin. When you install a Bitcoin wallet on your phone or computer, you are creating a private space that only you control. No company owns it. No one can open it without your permission. The wallet gives you a set of words known as a recovery phrase, and this phrase is the key to your money. Anyone who has those words can take everything in the wallet, so they must be kept offline and safe. Once the wallet is set up, you can receive Bitcoin from someone else in minutes, without filling out long forms or waiting for a bank to approve the transfer.

Sending Bitcoin works the same way. You type in the address of the person receiving it, confirm the amount, and the payment goes through. There are no banking hours, no delays, and no limits based on where the other person lives. Bitcoin moves any time of day, and it cannot be blocked by a company deciding to review your activity. This is one of the main reasons people use Bitcoin during emergencies. When systems freeze or websites crash, Bitcoin can move even when banks slow down or stop. This makes it a valuable backup during storms, political unrest, or financial disruptions. Women who manage households often understand the value of a backup system better than most, and Bitcoin offers one that does not depend on a single company or government.

But using Bitcoin safely takes awareness. Many people lose money not because Bitcoin is unsafe, but because they rush, trust strangers online, or treat it like a simple app. Unlike a bank account, Bitcoin has no "forgot password" button and no customer service line to call. If someone tricks you into sending Bitcoin, there is no way to reverse it. If you lose your recovery phrase, no one can recreate it for you. This is why scammers often target new users and why the loud online culture around Bitcoin can be dangerous. Women entering the space need calm, practical guidance, not intimidation or pressure from people who claim to be experts. The safest way to learn is slowly, with small amounts of money and a focus on understanding the basics before doing anything big.

Bitcoin is also public in a way that surprises many people. The network records every transaction on a shared ledger. It does not show names, but it does show addresses and amounts. If someone connects your address to your identity through a careless mistake, your transactions can become less private. This is why it is important to use a new address for each transaction and avoid posting wallet information publicly. Privacy with Bitcoin is possible, but it requires good habits. Women often handle sensitive tasks online, such as health planning, school communication, and household budgeting. Adding Bitcoin to their digital life means treating it with the same level of care.

Another thing women should understand is that Bitcoin moves differently from the money in a checking account. The value goes up and down, sometimes sharply. This does not mean Bitcoin is broken or failing. It is simply a new type of asset in a world where markets react quickly. Women who want stability do not need to invest large amounts. Many hold only small amounts as a form of emergency savings or as a way to protect a portion of their money from bank outages or financial restrictions. Bitcoin can be a shield even when the amount is small. The goal is not to gamble or chase large profits. The goal is to create a small corner of financial independence.

Above all, Bitcoin is most powerful when it gives women choice. A woman might keep most of her money in traditional accounts while holding a small amount of Bitcoin for privacy or protection. Another woman might use Bitcoin to send money to a friend in need without delays. Someone else might use it to set aside funds she alone controls. None of these choices require joining a culture or adopting a certain political view. Bitcoin becomes useful when it serves real life, not when it becomes a lifestyle.

The next section will explain how to avoid common traps, how to choose a wallet safely, how to protect your recovery phrase, and how to decide whether Bitcoin should be part of your financial plan. It will also include simple step-by-step guidance written for women who want clarity, not chaos.

Bitcoin becomes useful only when a woman understands how to keep it safe. The most important part of using Bitcoin is protecting the recovery phrase that the wallet gives you. These words are the only way to open the wallet again if your phone breaks or is lost. They should never be typed into email, saved in a phone note, stored in photos, or shared with anyone. The safest approach is to write the words on paper and store them somewhere secure, such as a fireproof box or a personal safe. This single habit protects more women from scams and loss than any other step. When the phrase is kept offline, no one can access your money even if they break into your accounts or steal your phone.

Choosing the right type of wallet is also important. Some wallets are simple apps that work well for small amounts, while others are physical devices that look like USB sticks and provide stronger protection for savings. Many women start with a phone wallet to learn the basics and then move to a hardware wallet once they understand how Bitcoin works. The key is not the brand or the style but the level of control. You want a wallet where you hold the recovery phrase yourself, because this means the wallet is truly yours. When a company controls your Bitcoin for you, it becomes more like a bank account than a private tool, and the company can freeze, limit, or lose your funds if something goes wrong.

Women should also know how to avoid the traps that cause most beginners to lose money. Many scams involve someone offering to "help" set up a wallet, match your investment, or manage your coins for you. These offers are always dangerous. No one trustworthy needs your recovery phrase, and no real opportunity requires you to send Bitcoin to a stranger. Another trap is rushing into trades or buying more Bitcoin than you understand. There is no need to move quickly. A small amount is enough to learn. Bitcoin does not require risk to be useful. It requires steady habits and calm decisions.

It is also helpful for women to understand that Bitcoin is not meant to replace every financial tool. It works best as part of a larger plan. Some women use Bitcoin as a private emergency fund that only they control. Others use it for long-term savings or for protection during unstable political moments. Some use it to support friends or family safely when traditional systems are blocked or slow. Many women keep most of their money in regular accounts and hold only a small amount of Bitcoin as a backup. This approach is practical and avoids stress. The value of Bitcoin comes from independence, not from trying to get rich.

Understanding Bitcoin also means understanding when not to use it. If someone is in danger, facing financial abuse, or trying to create distance from a harmful partner, Bitcoin must be handled carefully. It can provide safety because it is private and cannot be frozen by a bank, but it can also create risk if the recovery phrase is discovered by someone else. Women in these situations often use a separate device, a private email, and a hidden location to store their phrase. These small steps create protection that cannot be undone by outside forces.

When used with intention, Bitcoin becomes more than a trend. It becomes a tool for personal independence. It gives women a way to hold money safely, move it freely, and build a financial life that is not at the mercy of failing systems or unpredictable institutions. It works quietly in the background and waits for moments when it is needed most. That is what makes it powerful. It is not about noise, culture, or hype. It is about choice.

The next chapter, centered on Nostr, will show how communication can be reclaimed in the same way. It explains how women can speak, build community, and create identity on a platform that does not censor their voices or mine their lives for profit. It also shows how to move through that space with privacy, control, and the confidence that your words cannot be quietly filtered or erased.

Step-by-Step: Wallet → Send → Receive

Setting up a wallet:
Download a trusted Bitcoin wallet app from the official website or app store. When you open it, the app will give you a recovery phrase. Write the words on paper and store the paper somewhere safe. Do not take a photo of the phrase or type it into your phone. Once the recovery phrase is saved, the wallet is ready to use.

Receiving Bitcoin:
Open the wallet and tap "Receive." The app will show you a long address made of letters and numbers. This is like your bank account number but only for Bitcoin. Copy the address and give it to the person who is sending you Bitcoin. They will use it to send money to your wallet. You will see the funds in your wallet once the network processes the transaction.

Sending Bitcoin:
Tap "Send" in your wallet. Paste the address of the person you want to pay. Check the address carefully before confirming. Type in the amount you want to send, review the details, and approve the transaction. The payment will move directly to their wallet without a bank or company controlling the process.

This is all most users ever need to know. Once you learn these steps, you can handle the basics of Bitcoin with confidence.

Five Scams Women Should Avoid

Anyone asking for your recovery phrase.
No real person, company, or support team ever needs these words. Anyone who asks is trying to steal your Bitcoin.

People offering to "double" your money.
Bitcoin cannot be doubled by magic. Anyone who promises fast gains or special opportunities is setting a trap.

Fake customer support accounts.
Scammers often pretend to be support agents. Real support will never contact you first and will never ask for personal information.

Investment groups or pressure tactics.
If someone tries to rush you into buying Bitcoin or joining a so-called expert group, step back. Pressure is a warning sign.

Strangers asking for "help" with their wallet.
They often claim their account is locked or they cannot access their funds. They want you to reveal information that lets them steal from you.

These five scams cause most beginner losses. Knowing them protects you more than any technical skill.

Bitcoin as an Emergency Cash Bucket

Bitcoin can act as a small, private emergency fund that does not depend on a bank or a partner and cannot be quietly frozen or monitored. Many women use it to hold a small amount of savings that they can reach at any time, even during bank outages, blocked accounts, or personal emergencies. A Bitcoin wallet can stay silent for months or years and be opened only when needed. It does not ask questions, require documents, or follow office hours.

This makes Bitcoin useful during unexpected events such as job loss, relationship breakdowns, medical emergencies, travel problems, or times when regular systems are slow or unreliable. A woman does not need to hold a large amount. Even a small reserve can create real security. The goal is not to invest heavily or take risks. It is to build a private cushion that gives her control when other systems fail.

Bonus Tip: Never Use Leverage or Interest-Bearing Accounts

Leverage and interest-bearing crypto accounts are two of the fastest ways beginners lose money. Leverage means borrowing extra money to make a bigger bet, and it can erase your entire balance in minutes if the price moves the wrong way. Interest-bearing accounts are no better. They require you to hand your Bitcoin to a company that promises returns, but once they hold your coins, they control them. Many of these companies freeze withdrawals, collapse, or disappear during market stress. The safest path is simple. Hold your own Bitcoin, avoid borrowing to buy more, and keep your money in a wallet that only you control.

~4
Nostr: Speech, Identity, & Escape Routes

Most people assume online speech is free because the internet looks open and full of choices. In reality, almost every major platform is controlled by a company that decides what can be said, who gets seen, and which voices are allowed to grow. These decisions are shaped by profit, reputation management, and government pressure. Even when the rules are hidden, the effects are not. Women feel this control early and often. Posts about safety, harassment, inequality, or lived experience are flagged or buried because companies want to avoid anything considered "sensitive." When your daily reality is labeled a risk, your voice becomes easy to silence.

Nostr was created to solve this problem. It is not a company and does not belong to one group. It is a simple system that lets people publish messages that cannot be quietly removed by a platform. Instead of uploading your words to a single website, Nostr spreads them across many independent servers called relays. No single relay can silence you because many others can carry your words. This does not mean every message will go viral or reach everyone. It simply means your speech cannot be quietly deleted by a company that finds it inconvenient.

For women, this matters. Many women have learned to stay quiet online because speaking honestly leads to punishment. Reporting abuse, calling out injustice, or naming harm often triggers automatic moderation systems. These systems claim to protect the community, but in practice they hide the experiences women need to share. Nostr offers a different environment. It has no central authority and no algorithm pushing you down for being "too emotional," "too political," or "too real." It allows women to speak in their own voices without fear that a company will lock their account or shadow-ban their posts.

But Nostr is not perfect. It is not designed to make you famous or protect you from all harassment. It is designed to give you control over your words. Because there is no company running it, there is also no company to clean up the space. You choose who to follow, which relays to use, and how public you want to be. This freedom is powerful, but it requires steady habits. Women who use Nostr need to understand how to protect their privacy, how to move through the space confidently, and how to avoid the same traps that exist across the internet.

The purpose of this chapter is not to convince women to join Nostr. It is to explain what it is, why it matters, and how it can be used as a safe communication tool when other platforms close up or turn against you. The next section will explain how Nostr actually works, how identity is created on the platform, and how women can build safer spaces through careful choice of relays, profiles, and followers.

Nostr seems complicated at first because it works differently from the platforms most people know. Traditional platforms act like gated communities. You enter through one door, the company watches everything you do inside, and the company decides what is allowed, what is forbidden, and what will be seen. Nostr removes the gate. Instead of one door, there are many small doors called relays. When you post something, you can send it to one relay, ten relays, or a hundred. Each relay is run by different people, and none of them have full control over your identity or your words. This means you are no longer at the mercy of a single company or a single set of hidden rules.

Identity on Nostr works in an equally simple way. Instead of signing up with an email address and a password, you create a public key and a private key. The public key is like your username. The private key is what proves the messages are yours. No one else can post under your name if they do not have it. This also means you must protect the private key the same way you protect a password to your bank. Anyone who has it can take over your identity. Many women solve this by writing the key on paper and storing it in a safe spot instead of keeping it in their phone. This protects your account even if someone goes through your device or tries to break in digitally.

Because Nostr is open and not controlled by one company, women gain more control over their visibility. On most platforms, your posts live or die based on algorithms you never see. A system decides whether you look informative, risky, boring, emotional, or controversial. On Nostr, nothing pushes you down or hides your voice. If you publish a note, it stays published. If someone wants to see you, they can. The reach you gain comes from real connection, not from invisible rules. This can be freeing for women who are tired of being filtered, punished, or misread. It can also be grounding because it creates a sense of honest presence rather than competition.

Choosing relays is one of the most important parts of using Nostr wisely. Some relays are fast and public. Some are private or focus on certain communities. Some have basic moderation rules, while others allow anything. Women often prefer a mix. A few trustworthy relays keep your messages stable and easy to find, while a smaller number of private or women-led relays create a safer space for community conversations. Because you can leave a relay at any time, no relay owns your identity. If a relay becomes hostile or fills with harassment, you can disconnect from it without deleting your account or losing your history. This level of mobility is rare online, and it gives women power they do not get on centralized platforms.

Followers on Nostr also work differently. There is no algorithm suggesting who you should follow or pushing strangers into your feed. You decide whose voices matter. You decide whose notes you want to see. This allows women to build calmer, more intentional spaces. It also prevents the emotional exhaustion that happens on larger platforms where the feed is full of conflict, news cycles, and aggressive strangers. Nostr gives women the option to shape their environment instead of surviving inside someone else's design. That alone can feel like a quiet form of liberation. Safety on Nostr depends on habits rather than platform rules. Because the system is open, women must choose what they share and what they keep private. Many women use Nostr under pen names or partial identities that let them speak freely without exposing their personal lives. Others keep two separate profiles, one for public speech and another for community conversations. Some women post only from their computers to reduce the amount of location data tied to their phones. These habits do not create fear. They create space. They allow women to speak clearly without offering themselves up to systems that track every detail for profit.

Nostr is not a cure-all, but it is a tool that restores something women have lost online: the ability to speak without permission. When the goal is independence, clarity, or connection, Nostr gives women a place to build identity without constant surveillance or punishment. Setting up a Nostr account is easier than it sounds. The most important part is creating your keys, which act as your identity on the network. When you download a Nostr app, it will give you a long private key and a long public key. The public key is safe to share. The private key must be protected the same way you would protect your wallet password or your Bitcoin recovery phrase. Write the private key on paper and store it somewhere safe. When you keep the key offline, no one can take your identity or speak in your name.

Once your keys are safe, you can choose your relays. Relays are the places that store and share your posts. Most Nostr apps come with a few relays already added, but you can add more at any time. Many women start with a small group of stable relays so their messages are always available, and then add a few relays for quieter or more supportive communities. You can remove relays that feel hostile or chaotic and add new ones as your needs change. Instead of being trapped inside one company's rules, you shape your own digital environment.

Posting on Nostr feels simple because you write a note and send it to your relays. What makes Nostr different is the sense of ownership. Your posts are not filtered by a company trying to avoid controversy or protect advertisers. They stay up because you published them, not because an algorithm approved them. This creates steady ground for women who are tired of having their voices hidden. You choose what to say and who should see it, and that choice stays yours.

Even with this freedom, privacy still matters. Many women use partial identities so they can speak openly without inviting unwanted attention. Some keep their location private by avoiding posts that reveal routines. Others run two profiles with different purposes. These habits do not limit expression. They protect it. Nostr gives you room to be honest, but honesty does not require full exposure.

The most powerful part of Nostr is that you do not need permission to exist online. Most platforms control women's speech through rules that punish emotion, vulnerability, or uncomfortable truth. Nostr breaks that pattern. It allows women to speak, build community, and share knowledge without a system quietly reducing their reach. As with AI and Bitcoin, the goal is not to master everything at once. It is to take steady steps toward tools that strengthen your voice instead of weakening it.

Exact 10-Minute Setup

Minute 1–2:
Download a trusted Nostr app such as Damus (iPhone), Amethyst (Android), or a web client like Coracle. Open the app and choose "Create Account."

Minute 3–4:
The app will generate two keys: a public key and a private key. Copy the private key onto paper. Store it in a safe place. Do not put it in your phone notes, email, screenshots, or cloud storage.

Minute 5–6:
Add a simple display name and a short bio. You do not need a real name. A pen name or initials are enough.

Minute 7–8:
Check your default relays. Keep the ones that look stable. Add two or three more from the lists below.

Minute 9–10:
Make your first small post to confirm everything works. You are now live on Nostr, with full control over your identity and speech.

Starter Relay Lists

These relays give women a stable, safe, and calm experience on Nostr. You only need a few. Each one helps your posts stay visible without exposing you to chaotic or hostile spaces.

General, stable relays:
- damus.io
- nos.lol
- nostr.band
- primal.net
- nostr.land
- wellorder-relay
- creatr.nostr.wine

These relays offer strong uptime, wide reach, and a steady environment for everyday posting. They do not rely on one company and they do not filter your speech.

Optional quieter relays:
- nostr.mom
- nostr.wine
- relay.current.fyi

These relays move slower and feel calmer. They are good for women who want a quieter feed or smaller community spaces.

Note: You can remove relays whenever you want. You can add more later. No relay owns your identity or controls your account. You control your environment, your voice, and your visibility.

Engagement Plan for Non-Tech Creators

Non-technical women often grow faster on Nostr because the platform rewards honesty and steadiness, not technical jargon. Begin by posting simple updates or short reflections rather than perfect content. Respond to people who follow you, follow writers whose tone feels respectful, and share notes consistently. Nostr rewards presence, not performance. You do not need graphics, large threads, or deep technical knowledge. Clear writing and regular activity build trust and visibility.

If you create art, fiction, guides, or commentary, share small pieces of your work. If you run a project or platform, post gentle updates rather than marketing language. The goal is to show your voice, not to brand yourself. Nostr's culture values sincerity far more than polish.

How to Avoid the Male-Bro Swamp

Nostr is open, which means anyone can show up, including the loud, competitive male crowd that dominates many tech spaces. You do not need to engage with them. You can avoid the "bro swamp" through simple habits. Follow women, artists, writers, mothers, activists, and quiet creators. Disconnect from relays filled with shouting or hostility. Block or mute people who drain your energy. Build your feed intentionally so your space feels calm and human.

Most importantly, do not let anyone pressure you into debates, technical arguments, or ideological fights. Nostr gives you the power to create your own environment. Use that control. Women thrive on Nostr when they treat it like a room they decorate, not a room they must survive.

Bonus Tip: Protect Your Voice, and Feel Free to Explore

Your private key is the heart of your Nostr identity. Anyone who has it can speak as you or try to control your space, so keep it offline and safe. When you guard your key, you protect your freedom to speak without fear of being erased or silenced. After that, feel free to explore. Try new relays, test different profiles, and learn at your own pace. Nostr is a space where you can move slowly, experiment, and discover what works for you. The more you play with the tools, the more confident and independent you become.

~5
Digital Self Defense for Women

Most people carry their entire lives in their phones without realizing how exposed that makes them. A phone holds messages, photos, location history, contacts, passwords, health details, and personal routines. It also keeps records of where you go, who you talk to, and what you search for. For women, this creates real vulnerability. A phone can protect you, but it can also reveal parts of your life you never meant to share if it is not secured. Digital self-defense starts with one simple truth: your phone is not just a tool. It is a record of your life, and you have the right to control who sees it.

Modern systems collect data constantly. Many apps track location even when you are not using them. Some read your contacts, scan your photos, or monitor what you search for. Most companies claim this is harmless or helpful, but the real purpose is to build a profile of your habits. These profiles can be used for ads, but they can also be shared, sold, or analyzed by systems you never see. Women who already navigate work, caregiving, relationships, and safety concerns do not need their digital footprint working against them. Reducing that footprint is one of the most powerful steps a woman can take to protect her privacy and independence.

The goal of digital self-defense is not to live in fear or avoid technology. The goal is to take back control.

When you understand how your phone collects information, you can make choices that limit what companies learn about you. You can decide which apps deserve access to your location and which ones do not. You can stop giving away personal details to systems that do not protect you. You can build a digital life that supports you instead of tracking you.

Women face specific risks because their digital devices often include details about children, medical care, work schedules, and caregiving duties. This means a woman's phone often contains information about other people she is responsible for, not just herself. Even something small, like a shared calendar or a school portal, creates a trail that companies or automated systems can read. When a woman strengthens her digital privacy, she protects not only her own safety but also the people who depend on her. Digital self-defense becomes a form of real-world protection.

Phones also create emotional exposure. Old messages, private photos, and sensitive conversations can stay stored for years. They may feel invisible to you, but they are not invisible to the device. Anyone who gets into your phone, even briefly, can read far more than you expect. That includes partners, strangers, coworkers, and automated scanning systems. This does not mean you must delete everything. It means you should understand which parts of your digital life need more protection. A simple passcode, stronger settings, and better habits can remove most of the risk without changing how you use your phone day to day.

Part of building digital strength is learning to slow down. Many women keep their phones unlocked or use simple passcodes because life moves fast. They also download apps quickly without checking what those apps can access. Digital companies rely on this. They want you moving too fast to notice what they collect. When you pause before tapping "allow," you take back power. When you set boundaries on what your phone can share, you make it harder for companies, platforms, and bad actors to learn more about you than they should.

This chapter will show you how to harden your accounts, shrink your digital footprint, and protect your privacy without changing your personality or your routine. It does not matter if you are tech-savvy or overwhelmed. Most of these steps take a few minutes and require no technical background. What matters is consistency. When you build small habits that guard your information, you create a safer digital life that lasts.

Securing your phone begins with securing your accounts, because most companies treat your phone number and email address as proof that you are really you. If either one is weak, everything connected to it becomes weak. The safest step is to use strong passcodes and turn on two-factor authentication for your essential accounts. This includes your email, your phone account, your cloud storage, and any service that holds personal information. A strong passcode does not need to be complicated. It just needs to be something no one around you can guess. This single change is one of the fastest ways to block someone from breaking into your digital life.

Location tracking is the second major threat for women. Phones store where you sleep, where you shop, where your children go to school, and which routes you take daily. Many women do not realize how many apps have had long-term access to their location. Most apps do not need this information to function, and many use it to build profiles for advertising or analysis. Turning off location services for anything that does not truly require it shrinks your digital footprint immediately. You can also choose to share location only when the app is open, which keeps your movements private the rest of the time.

Your phone also reveals information through photos, Bluetooth, Wi-Fi networks, and automatic backups. Photos often include hidden data about where and when they were taken. Bluetooth can connect to devices you do not expect. Wi-Fi networks can track when you come and go. Backups can store private details in places you rarely check. None of these facts mean you should stop using your phone. They simply mean you should understand how easily small details can be collected. A few minutes adjusting your settings can block most of this silent tracking and help you keep your routines out of systems that do not deserve that level of insight.

Women also benefit from limiting the number of apps on their phones. Every app is another doorway into your information. Many apps collect far more data than they need because the business model depends on it. Before installing anything new, ask yourself what the app offers and whether it truly requires the access it requests. If it wants your contacts, your microphone, your photos, or your exact location without a clear reason, choose something else. Deleting unused apps helps too. Apps you forgot about can continue collecting data in the background, even when you do not interact with them.

Messaging habits matter as well. Many women rely on group chats, family threads, workplace channels, and school apps, and each one creates a different layer of exposure. Not all messaging apps protect your privacy equally. Some scan message content, while others store long histories in the cloud. Choosing a messaging app that respects privacy gives you more control over what stays on your device and what leaves it. Even small habits, such as deleting old conversations or turning off cloud backups for chats, can reduce exposure without changing how you communicate.

Pseudonyms are another tool that many women find useful. A pseudonym gives you space to speak honestly without tying your words to your full identity. This does not mean living anonymously or hiding from the world. It means creating a boundary between your public life and your private one. Using a different email, a separate account, or a name that protects your personal details can offer safety when discussing sensitive topics, participating in activism, or navigating spaces where women are often targeted. A pseudonym works best when it is used consistently and when you keep identifying details out of the profile. It becomes a protective layer that lets you communicate without exposing parts of your life that should remain yours alone.

The final part of this chapter will bring everything together. It will guide you through the full "20-minute lockdown," give you an iPhone and Android settings checklist, explain pseudonym best practices in simple steps, and outline what women should avoid posting online to stay safe. These tools will help you build a digital environment that supports your well-being without requiring you to become an expert.

Protecting your digital life does not require hours of effort. Most women can secure their phones, accounts, and identities in a short, focused window. A "20-minute lockdown" is a fast routine that strengthens the weakest points of your digital world. Start with your lock screen. Use a strong passcode and turn off any features that show messages or private content while your phone is locked. Next, check your main accounts and turn on two-factor authentication. These steps stop most intrusions before they begin. Then move to your location settings and turn off access for apps that do not need it. Finally, delete apps you no longer use. By the time twenty minutes have passed, your phone becomes harder to track, harder to enter, and far less exposed.

Securing your settings is the next step. iPhones and Android phones both offer tools that make your data harder to access, but these tools are often turned off by default. On either device, you can limit how often apps check your location, stop apps from accessing your microphone and camera, and control which apps can see your photos. You can also check your privacy dashboards to see what information apps have used recently. These settings may feel hidden, but they are quick to adjust once you know where to look. Many women are surprised by how many apps had long-term access to sensitive information. Changing these permissions shrinks your digital footprint immediately.

Pseudonyms work best when they create separation between your real identity and your public presence. A strong pseudonym uses a fresh email address, a separate profile photo, and no identifying details. This means no birthdays, no home locations, no children's names, and no workplace references. A good pseudonym lets you speak freely without tying your words to your private life. It also protects you from people who may search your history to gather information about you. When used consistently and carefully, a pseudonym becomes a shield that allows women to take part in conversations without giving strangers a pathway into their lives.

Knowing what not to post is just as important as knowing how to secure your phone. Anything that reveals your location, your routine, or your vulnerable moments can become a risk. Avoid posting where you live, where you work, or where your children go to school. Avoid sharing real-time locations, especially when you are alone or traveling. Do not post documents, IDs, medical notes, or screenshots that contain personal details, even if they seem harmless. And avoid long emotional posts that reveal patterns in your life. You do not need to hide who you are, but you also do not need to give the internet a map of your private world.

Digital self-defense is not about fear. It is about ownership. Taking control of your settings, your accounts, and your identity gives you freedom to use your phone with confidence. The more you build these habits, the safer your digital life becomes. None of this requires deep technical knowledge. It only requires attention, awareness, and a willingness to protect the parts of your life that should remain yours alone.

This chapter closes the loop on personal digital protection. The next chapter brings these tools together into a full safety net, one that combines private communication, financial independence, and digital resilience. With these pieces in place, women can move through the modern world with more strength, more clarity, and far greater control than the systems around them ever expected them to have.

The 20-Minute Lockdown

Minutes 1–3:
Set a strong phone passcode. Turn off lock-screen previews for messages and notifications.

Minutes 4–6:
Turn on two-factor authentication for your main accounts: email, phone carrier, cloud storage.

Minutes 7–10:
Open your location settings. Turn off location access for all apps that do not need it. Set the rest to "While Using."

Minutes 11–14:
Delete apps you do not use. Each deleted app removes a doorway into your data.

Minutes 15–18:
Check your privacy dashboards. See which apps used your camera, microphone, location, and photos. Remove permissions that feel unnecessary.

Minutes 19–20:
Restart your phone. This clears temporary data and applies your new privacy settings.

iPhone / Android Settings Checklist

For both:
- Strong passcode
- Two-factor authentication turned on
- Location set to "While Using" or off
- Microphone and camera permissions reviewed
- Photo access limited to selected images
- Bluetooth off when not needed
- Wi-Fi auto-join turned off for unknown networks
- App tracking blocked
- Cloud backup settings reviewed
- Privacy dashboards checked weekly

iPhone-specific:
- Lock screen previews disabled
- Significant Locations turned off
- Ad tracking limited
- Siri data minimized
- iCloud backups checked for message history

Android-specific:
- Google Activity Controls trimmed
- Nearby Device scanning disabled
- Autofill data cleared
- Google location history paused
- Play Store permissions reviewed

Pseudonym Best Practices

• Use a separate email address created only for your pseudonym.
• Choose a name that reveals nothing about your real identity.
• Do not use real photos. Use art, symbols, or landscapes.
• Avoid sharing birthdays, workplaces, schools, or hometowns.
• Keep your writing style simple so it is harder to trace.
• Do not connect your pseudonym accounts to your real phone number.
• Store login information in a safe place, not in your main phone notes.
• Never mix your real name and pseudonym on the same platform.

What NOT to Post

• Your home address, workplace, or school locations
• Real-time updates of where you are
• Photos of children with identifying backgrounds
• Screenshots of medical records, IDs, tickets, or documents
• Travel plans before or during the trip
• Emotional posts that reveal patterns in your routine
• Anything showing your daily commute or predictable habits
• Financial screenshots, account balances, or receipts

Bonus Tip: Protect Yourself When Systems Fail You

Government systems collect huge amounts of information about women, but they are not designed with women's safety in mind. They track benefits, addresses, school records, medical care, and past decisions, and they often share this data across agencies without asking you. When these systems make mistakes, they rarely fix them quickly. When they misuse information, women pay the price first. You cannot control what a government chooses to store, but you can control what you share. Keep your private life off platforms that feed government databases. Limit what goes into apps connected to health care, schools, or public services. The less they know, the less they can misread or misuse. Protect your information because no system protects women as well as women protect themselves.

~6
Building Parallel Power

Building parallel power means creating parts of your life that do not depend on systems that misunderstand, limit, or punish women. It means building income, communication, and identity in places that cannot be quietly shut down by a boss, a company, a platform, or a government. Parallel power is not rebellion. It is preparation. It is the steady work of building options that belong only to you, so your survival never rests on one fragile system or one person. When women build parallel power, they gain freedom from the forces that try to control their choices.

Parallel power starts with the idea that women deserve more than one channel for income and expression. Many women rely on a single job, a single account, or a single platform. That makes life brittle. If anything collapses, a layoff, an illness, a policy change, a platform shutdown, the entire structure of their life shakes with it. Building even one extra outlet, whether it is a small micro-business, a digital shop, a writing identity, or a skill-based service, gives women an anchor that cannot be taken away as easily. One small independent stream can change everything, even if it feels tiny at first.

Women often think building a business requires money, skill, or a big audience. It does not. A micro-business is simply a tiny, repeatable way to earn income on your own terms. It can be as small as selling digital products, posting guides, offering simple services, writing under a pseudonym, or sharing creative work in a form others want to buy. The point is not to become wealthy. The point is to build a corner of financial life that belongs to you. A space where you choose the work, the timing, and the boundaries.

Anonymous publishing is one of the simplest ways women can create parallel power today. You can write, post, teach, or sell without revealing your real name or your full identity. This protects your future job, your personal life, and your safety while giving you a way to speak honestly and create income. An anonymous identity can produce small guides, digital books, print-on-demand products, or quiet commentary. Many women find it freeing to build something without the pressure of being seen. Anonymous work becomes a shield that lets you explore your voice without fear of surveillance or judgment.

Communication is another pillar of parallel power. When your speech lives on a platform that can remove you, you do not have power. When your followers, readers, or supporters can only reach you through one corporate feed, you do not have power. Parallel communication means spreading your voice across multiple channels, especially those that do not rely on central control. It can be as simple as a backup email list, a Nostr account, a small website, or a private chat space. These small channels give you a way to stay connected when larger systems change their rules, hide your work, or silence your speech.

Parallel identity is a final piece that many women overlook. You do not have to live with only one identity online. A pseudonym, a pen name, or a project name can become a separate form of self that is safer, freer, and better suited for work, art, or commentary. When your identity is not tied to your real name, employers and institutions lose their power over your expression. You gain room to experiment and create without risking your personal life. Women have used alternate identities for centuries to escape judgment, to create opportunities, and to protect themselves from systems that do not care about their truth.

We've set the foundation for what parallel power means and why it matters now. Section two will show how women can start building these parallel channels through simple steps: opening small income streams, setting up anonymous publishing identities, choosing independent communication tools, and structuring micro-businesses that do not require high skills or large budgets.

Finally, section three will bring it all together with a full three-stream income plan, anonymous micro-brand templates, and a simple playbook for reaching people without using platforms that can erase you.

Parallel power grows through small, simple steps that build on each other. Many women think they need a full plan before they begin, but the truth is that most parallel systems start quietly. You do not need a perfect idea or a large amount of time. You only need one small action that creates a little more independence in your life. Once you take that first step, the next one becomes easier. Each piece you build becomes a part of a larger structure that belongs only to you.

The simplest place to start is with income. A small income stream gives women a buffer against sudden changes in their lives. You do not need a big project or a large audience to begin. A single skill, interest, or routine can turn into a micro-business. Some women create digital guides, small eBooks, or simple services. Others share tutorials, printables, or creative work that people want to buy. The point is not to build an empire. The point is to create a little financial space that is not controlled by an employer, a platform, or a partner. Even a small income stream becomes a form of protection.

Anonymous publishing is often the easiest gateway into parallel income because it requires almost nothing to start. You can write under a name that protects your identity, use a separate email, and post your work without exposing your personal life. This gives women the freedom to explore ideas, test products, share knowledge, or build a voice without fear of being judged or monitored. You can publish on platforms that do not require your real name and offer digital products that take only a few hours to create. You do not owe your personal identity to the internet. You can create from a safe distance.

Communication is the next step. Women need more than one channel to reach people. Relying on a single platform is fragile because platforms change their rules without warning. A small alternative channel can be as simple as a private mailing list, a Nostr account, or a small website with one page and a contact form. These channels do not need to be active every day. They simply need to exist so you have a way to speak, publish, or update people even if a larger platform turns against you. Communication becomes part of your safety net.

Women can also build parallel power by creating small, independent identities for different parts of their lives. A pseudonym for writing. A project name for selling digital goods. A separate identity for activism or community work. These identities do not replace your real self. They protect your real self. They let you explore different forms of expression without tying everything back to your personal life. Many women feel more confident and creative under a pseudonym because it removes the pressure and fear that often come with being fully visible.

Parallel power also includes small technical choices that give you more control. Setting up a private email that is not tied to your main accounts gives you a clean space for business or anonymous work. Using simple tools that preserve privacy protects your projects from platforms that track and profile users. Even switching to payment options that allow digital products without revealing your identity can strengthen your independence. These changes do not require technical skill. They only require intention.

This section covered how to build the pieces. Next are the steps for organizing them into a full plan, how women can set up a three-stream income system, create anonymous micro-brands, and use a simple distribution playbook that does not rely on platforms that can silence or punish them. Together, these tools form a complete, durable system of independence.

Three-Stream Income Plan

A strong parallel income system does not start with a big business. It starts with three small, simple streams that support each other.

Stream 1: A quick digital product
This can be a guide, checklist, template set, short workbook, or small eBook. It requires almost no cost and can be created in a few hours. It gives you something you own, something you can sell, and something that does not vanish if a job or platform disappears.

Stream 2: A tiny service you can offer
Choose something you already know how to do. This could be editing, formatting, simple design, tutoring, organizing, researching, or giving feedback on projects. A tiny service turns your time into direct income and gives you a steady backup when other systems fail.

Stream 3: A slow, long-term skill or brand
This is the identity you build over time. It can be your pseudonym, your micro-brand, a small newsletter, a creative project, or a quiet Nostr presence. This stream grows slowly but becomes the most stable one over time. It gives you a place to express yourself without relying on institutions.

When you have all three, you have a safety net. One brings quick wins, one brings steady options, and one builds long-term independence.

Templates for Anonymous Micro-Brands

A micro-brand does not need a real name, a large audience, or a perfect plan. It only needs clarity and safety.

Template A: Skills-Based Identity
Name: A simple project name or pseudonym
Focus: One skill (editing, art, guides, templates)
Tone: Helpful, calm, steady
Output: Small digital products + low-time services

Template B: Creative or Writing Identity
Name: Pen name that protects your real life
Focus: Commentary, fiction, guides, or essays
Tone: Clear, grounded, reflective
Output: Short eBooks, posts, audio notes, print-on-demand

Template C: Survival / Practical Identity
Name: Functional or symbolic pseudonym
Focus: Checklists, safety guides, routines, simple tools
Tone: Direct, minimal, practical
Output: Micro-manuals, planners, worksheets, tip sheets

Each micro-brand should use:
• a separate email
• a simple profile photo (symbol, art, object)
• no personal details
• one place to publish and one place to sell

This keeps the brand safe, clean, and easy to grow.

Simple Distribution Playbook

You do not need algorithms or large platforms. You need three steady paths for your work:

Path 1: A platform that cannot erase you
Use Nostr as a base. Post small updates, previews, or notes. This gives you a permanent record that companies cannot hide or remove.

Path 2: A place people can buy or download
Use a digital storefront or paywall you control. Gumroad works well for small creators, especially under pseudonyms. Your products stay available even if social platforms change their rules.

Path 3: A backup channel you own
Have a simple mailing list or one-page website. It does not have to be fancy. It only needs a signup form or a link to your work. This keeps your audience connected no matter what happens elsewhere.

These three paths make your work reachable, even when bigger systems fail or turn hostile.

Bonus Tip: Build Quietly and Without Permission

You do not need approval to create your own streams of income, your own identity, or your own communication channels. Systems that were not built for women will never give you a perfect moment to start. Build quietly. Build slowly. Build in ways that protect your privacy and your safety. You do not need to announce your projects or justify them to anyone. Parallel power grows strongest when it grows on your terms. Even the smallest effort becomes a form of freedom when it belongs only to you.

~7
Crisis Mode: When Systems Start Closing

Most people imagine "crisis" as something dramatic, sirens, headlines, sudden shocks, when in reality, systems usually close slowly. The warning signs appear in small changes that are easy to overlook. A policy update here. A new requirement there. A platform tightening its rules. A bank freezing accounts during "unusual activity." A service demanding more identification than before. Each step feels minor on its own. Together, they create a world where women have less room to move, fewer choices, and less control over their own information.

For women, these quiet closures show up early. Women rely on digital systems for work, caregiving, school communication, health appointments, and basic safety. When institutions change their rules, women feel the pressure first. A school platform suddenly restricts messages. A healthcare portal locks information behind new verification steps. A bank flags deposits or freezes transfers during "system checks." A social platform removes posts about danger or inequality. None of these moments look like crisis on the surface, but they limit access in ways that matter. When a system stops working for women, the consequences are immediate.

Crisis mode does not mean chaos. It means increased friction. When the tools you rely on become slower, stricter, or less predictable, it becomes harder to live your daily life. A woman trying to schedule a medical appointment may find that the system requires new logins or added documentation. A mother trying to communicate with a school may discover that her messages get filtered, delayed, or lost. A woman applying for a job may never learn that an automated system rejected her before a human ever saw her name. These moments create stress, confusion, and instability, all without public acknowledgement that anything is wrong.

When systems tighten, women often blame themselves instead of recognizing the pattern. They think they made a mistake, missed a step, or used the wrong setting. They assume the system is right and they are wrong. This is exactly how quiet control works. Systems appear neutral, but their design hides their impact. A woman trying to understand why her post was taken down or her account was flagged may never receive a real answer. She is left guessing, adjusting, and shrinking her presence to avoid being punished again. Crisis mode pushes women toward silence long before they recognize that silence as a response to pressure.

The key to navigating a closing system is learning to recognize the friction for what it is: a signal. Not a failure on your part, but a shift in how institutions operate. When communication becomes harder, when access becomes slower, when verification demands multiply, and when your voice becomes easier to suppress, the system is not losing capacity, it is tightening control. You cannot stop these shifts alone, but you can prepare for them. You can build alternate channels, protect your information, and create small structures that do not collapse when the bigger ones do.

Section one defines the terrain of crisis mode, slow tightening, quiet restrictions, and rising friction. Section two will show how to recognize these signs early and how women can respond with steady steps that keep them safe even as systems push them out. Section 3 will turn this into action with a practical guide for staying functional during institutional failures or digital lockouts.

Recognizing a closing system early is the difference between being overwhelmed and being prepared. Most people do not notice the shift because it happens in small steps. A form becomes harder to access. A phone system no longer connects you to a human. A login requires more identification. A website demands a new verification method. Systems rarely announce that they are tightening their controls. Instead, they ask for more from you each year, more data, more compliance, more patience, while giving less clarity in return. Women, who already carry more responsibility for navigating schools, healthcare, family accounts, and communication platforms, feel these shifts before anyone else.

The first sign of a closing system is *increased friction*. A task that was simple last year suddenly requires multiple steps. A service that used to respond quickly now takes days. A platform that once allowed open communication now removes posts without warning. Friction is not random. It is a signal that the system is prioritizing its own rules over your ability to function. Women often experience this as confusion or self-blame, wondering what they did wrong. But the truth is that the system has changed, not you.

The second sign is *rising verification*. Systems start demanding more identification than necessary: government-issued IDs, biometric checks, personal history questions, or repeated logins. What starts as "security" often becomes routine surveillance. Women should pay attention when everyday processes require information that feels excessive. When a school portal wants access to your phone number and location, when a healthcare system demands constant two-step verification, or when a bank requires documents you have never needed before, it means the system is tightening control over access.

The third sign is *shrinking communication channels*. Schools, companies, hospitals, and platforms increasingly limit how people can be reached or how problems can be reported. Instead of real people, women encounter automated chatbots, dead-end menus, or message systems that never respond. This shrinkage creates frustration, but more importantly, it creates silence. When a system can no longer be questioned, appealed to, or contacted, it becomes unaccountable. Women should treat disappearing communication as a warning, not an inconvenience.

The fourth sign is *quiet punishment*. This includes sudden account restrictions, automatic flags, message removals, frozen funds, or access denials without explanation. These penalties rarely come with a human explanation. Instead, an algorithm makes a judgment and the system enforces it without context. Women may assume they did something wrong, but usually the trigger was a pattern the system misread. Quiet punishment is one of the clearest indicators of authoritarian drift within institutions: decisions that impact your life but cannot be challenged.

The final sign is *dependence without alternatives*. When people rely entirely on one platform, one account, or one digital identity, the system gains full leverage. If it stops working, the person loses access to critical parts of their life. Women feel this sharply when a school communication app goes down, when a bank freezes a card, or when a healthcare portal blocks access to records. The more a system becomes unavoidable, the more dangerous its failures become. Recognizing this pattern early allows women to build alternatives before they are needed.

Understanding these signs does not require fear. It requires awareness. When women recognize friction, verification, communication shrinkage, quiet punishment, and forced dependence for what they are, patterns of tightening control, they can prepare without panic. They can create backup systems, duplicate important information, build separate communication channels, and protect their identity before the pressure increases. Crisis mode is not about waiting for collapse. It is about paying attention to the small shifts that reveal how institutions are changing around you.

This section explains how to see these changes clearly and respond early. The next, and final section, will give practical steps for staying functional during disruptions, lockouts, or system failures, and show how to protect yourself when institutional support becomes unreliable.

When systems tighten, the goal is not to fight them head-on but to stay functional, protected, and prepared. Crisis mode is not a single moment. It is a shift in how institutions behave, and women can meet that shift with practical actions that give them stability while everything around them becomes less reliable. These steps are not complicated. They are small habits that keep you from being knocked off balance when access breaks, accounts freeze, or communication collapses without warning.

The first step is *creating duplicates of the information you rely on*. This includes school contacts, medical details, medication lists, essential phone numbers, financial records, and household routines. When everything lives inside a single app or portal, the moment that system locks you out, you lose the ability to act. Keeping a simple offline copy moves the power back into your hands. A printed sheet, a written list, or a small secure file on a separate device is enough. These backups are not dramatic; they are practical. They let you keep moving even if the primary system goes dark.

The second step is *building backup communication paths*. When a platform restricts you, you should still have ways to reach the people and services you depend on. This can be a secondary email, a private messaging app, a Nostr account, or a simple text-based group. The point is not to use all of them daily. The point is to have somewhere to go when your main channel becomes unresponsive. Women often carry the emotional and logistical weight of communication for families, partners, and communities. A backup path prevents you from becoming isolated when a system stops listening.

The third step is *keeping a small amount of independent money available*. Even a few dollars outside a traditional bank, cash, Bitcoin held in your own wallet, or a prepaid card, can help you stay steady if your account is flagged or delayed. Financial lockouts tend to hit without warning. They may last hours or days. They may come from a glitch, a fraud check, or an automated mistake. Women who rely on a single account have no room to maneuver. A little independent money is not about wealth; it is about mobility.

The fourth step is *separating identities when needed*. Crisis mode is often the moment when women realize how exposed their single digital identity really is. If one account is compromised or blocked, every part of life connected to it becomes unstable. Having a simple pseudonym for online activity, a separate email for sensitive matters, or a second device for private tasks reduces this exposure. You do not need to hide who you are. You only need enough separation to stay safe when a system makes a decision about you that you cannot challenge.

The fifth step is *simplifying your digital footprint*. During system instability, the people who struggle most are those who depend heavily on complex digital tools. Women can protect themselves by reducing what they rely on. Fewer apps mean fewer points of failure. Fewer connected accounts mean fewer ways to be locked out. Simplifying your phone, your services, and your digital routines helps you stay in control when systems demand more identification, more confirmation, and more patience than before.

The final step is *learning to pause and read the pressure correctly*. Crisis mode often pushes women to rush, fix the error, send the document, upload the ID, answer the prompt. But rushing in response to pressure is how people give away information they did not mean to share. When a system becomes demanding or confusing, stop, breathe, and step back. Ask yourself whether the request makes sense or whether the system is tightening again. Women gain enormous protection by refusing to panic in the moment of the demand. A slow response often leads to safer choices.

This section brings your awareness into action. Crisis mode is not about fear or prediction. It is about having enough control and clarity to move through tightening systems without losing access to your life. When you keep copies of important information, build backup communication, hold a little independent money, create separation in your identity, simplify your digital footprint, and pause before reacting to pressure, you remain steady while the system becomes unstable.

Early Warning Signs of a Closing System

1. Increased friction
Tasks take longer, require more steps, or demand repeated attempts. Systems "glitch" more often and rarely explain why.

2. Rising verification demands
You are asked for more ID, more documents, or more personal information than before, even for simple tasks.

3. Shrinking communication channels
Support lines disappear. Messages go unanswered. You encounter bots instead of people.

4. Quiet punishment
Accounts freeze, posts vanish, transfers stall, or access is limited without any human explanation.

5. Forced dependence
You realize you rely on one platform, one device, one account, or one portal for essential tasks, and there is no alternative if it fails.

Red Flags Women Should Not Ignore

- A system locks you out without telling you why.
- You cannot reach a human to fix a problem.
- Your posts or messages are removed without clear reasoning.
- A bank, school, or medical portal demands new identification suddenly.
- Your account is flagged or restricted for the first time.
- You feel afraid to speak honestly online.
- You find yourself doing extra emotional labor to "fit" a system's rules.

Any one of these signals a shift in power, not a mistake on your part.

Backup Essentials to Prepare Before a Breakdown

Information copies:
school contacts
medical details and medication lists
essential phone numbers
financial account info
household routines or schedules

Communication backups:
a second email address
a private or alternate messaging app
a Nostr account or simple website
a pre-made group text for emergencies

Financial backups:
a small amount of cash
a small amount of self-custodied Bitcoin
a prepaid card for digital purchases

None of these need to be large. They need to be yours.

Crisis Mode: The One-Page Response Plan

If a system suddenly locks you out:
Stop and pause, do not rush to upload ID or more data.
Switch to your backup communication channel.
Access your offline copies of important information.
Use your independent money for immediate needs.
Document what happened for your own records.
Try again later using a different device, browser, or connection.

If a system becomes hostile or unpredictable:
Reduce what information you share with it.
Separate identities, use a backup email or pseudonym.
Shift essential tasks to slower, safer channels.
Begin building an alternative path permanently.

If multiple systems start tightening at once:
Focus on access to money, communication, and health.
Alert only trusted people to what is happening.
Switch to simplified routines that depend on fewer apps.
Move your digital footprint off high-risk platforms.

Crisis mode is not panic.
It is preparation and controlled movement.

~8
The EATMS 7 Day Starter Plan

Days 1–2: Stabilize Your Foundation

The best way to gain control in a tightening system is to begin with the parts of your digital life that expose you the most: your phone, your accounts, and your personal information. Days 1 and 2 focus on securing these foundations. You are not changing your entire routine. You are strengthening the base so that everything that comes later, income, identity, communication, stands on solid ground. When your phone, accounts, and basic tools are harder to track and harder to misuse, you regain a sense of stability that many women have not felt in years.

Day 1 begins with your phone, because your phone is the single most vulnerable point in your daily life. It holds your contacts, your messages, your photos, and the record of where you go each day. Securing it is not about perfection; it is about reducing exposure. Start by checking your passcode and turning off lock-screen previews. Review your location settings and remove access from any app that does not need it. Turn off Bluetooth when you are not using it. Delete any app you cannot explain or have not used in months. These small steps reduce the amount of data you leak without changing how you use the device.

Once your phone is safer, move to your main accounts. Your email, your cloud storage, and your phone carrier account are the keys to everything else. Turn on two-factor authentication, remove old connected devices, and change passwords that have not been updated in years. These steps protect you from lockouts, hacks, and automated flags that can interrupt your life without warning. Women often carry the emotional and practical burden of managing family accounts as well, which makes a strong foundation even more important. When your primary accounts are stable, you have more space to breathe.

Day 2 is about setting boundaries with artificial intelligence. AI now sits inside almost every digital tool, but women do not need to feed it personal information to benefit from what it can do. Keep your searches, planning tasks, and writing prompts separate from medical details, financial information, or private family matters. This is the time to create a secondary email or a separate login for AI use. If you want even stronger protection, use a different browser or a different device for tasks that involve sensitive topics. These boundaries protect you from becoming a data source that systems read, sort, and store without your consent.

The goal of Days 1 and 2 is not to overhaul your entire digital life. It is to create a base layer of safety and clarity you can build on. By the end of Day 2, your phone is cleaner, your accounts are protected, and your AI boundaries are in place. You may not feel dramatically different, but you will be standing on steadier ground. These small steps create a foundation that gives you more control when systems become unpredictable.

Days 3–5: Build Parallel Power

With your foundation secured, the next step is to build small, independent systems that do not depend on any platform, employer, institution, or algorithm. Days 3 through 5 focus on creating parallel power. These steps are simple, low-cost, and require no technical background. The goal is not to build a new life in one week. The goal is to build the structures that protect you when the larger systems around you tighten or fail.

Day 3 is about creating a small income foothold. Choose one simple digital product or service you can offer without stress. It should be something you already know how to do or something you can create quickly. This could be a short guide, a template set, a checklist, a printable, a one-page resource, or a tiny service like editing, reviewing, formatting, or organizing. The point is not to make a profit overnight. It is to create one small stream that belongs only to you. This parallel income will become a safety valve when other systems grow restrictive or unpredictable. Many women underestimate how much confidence and stability comes from earning even a few dollars independently.

Day 4 focuses on building an identity that protects your real life. Most women use one identity for everything, which makes them vulnerable when systems read, judge, or misinterpret their online presence. Creating a pseudonym or micro-brand gives you a safe space to publish, create, share, or experiment without exposing your personal information. Pick a simple name, create a separate email, choose a neutral profile image, and avoid tying it to your past. You are not hiding. You are separating your creative life from the systems that track your real one. Even if you do not publish anything yet, the identity itself becomes a tool you can use whenever you need it.

Day 5 is about building communication and financial independence. Set up one alternate communication channel: a secondary email, a private messaging app, or a Nostr account. You do not need to use it daily; you only need it to exist. This channel becomes your backup when platforms restrict your voice or when you need a space free from corporate or government filtering. Next, create a simple Bitcoin wallet, self-custody, no exchange storage. Even a small amount of independent money can help you stay steady during bank delays, financial flags, or digital lockouts. You are not investing. You are establishing mobility. A few dollars of cash, a few dollars of Bitcoin, and a backup card give you an immediate buffer that most women do not have.

By the end of Day 5, you have built three pillars of parallel power:
- a small income foothold
- a protected identity
- a backup communication and financial channel

None of these require skill or speed. They require consistency. These steps create room in your life, room to speak, room to earn, room to move, without waiting for permission from systems that were not made for you.

Days 6–7: Create Your Safety Net

The final two days focus on building the long-term structures that keep you steady when systems tighten again. You are not trying to overhaul your life. You are creating a safety net made of communication freedom, basic distribution, and small routines that protect your voice and your work. By this point, your foundation is secure and your parallel tools are in place. Now it is time to connect them so they support you even when the outside world becomes less stable.

Day 6 focuses on speech freedom. This does not mean sharing everything or speaking everywhere. It means building one space where you can communicate without being quietly filtered, punished, or erased. Set up your Nostr account fully, add a few stable relays, and make one or two simple posts. You do not need an audience. You need a place to speak that cannot be taken away by a company or algorithm. Many women feel a deep sense of relief the moment they have one corner of the internet that is not shaped by someone else's rules. Even if you post rarely, the account itself becomes part of your safety net.

Next, connect your parallel identity from Day 4 to your new communication space. Use your pseudonym, your alternate email, and your simple profile image. This keeps your real life protected and gives you a place to grow without pressure. The point of Day 6 is not performance. It is presence. When systems narrow the ways women can speak, having even one independent channel gives you freedom others cannot take.

Day 7 is about building a simple distribution routine. This is the step most women skip, but it is what turns small tools into real power. Choose one platform for your micro-brand: a Gumroad page, a tiny website, a simple blog, or a one-page link hub. Upload one small product, even if it is free. Write a short description. Add your alternate email. Link your Nostr account if you want. This becomes your distribution home, a place for people to find your work even when other platforms fail or push you out.

You do not need a schedule, a strategy, or a large audience. All you need is one place where your work can live without relying on a system that may censor or restrict you later. This single move protects you from losing everything when a platform changes its rules or decides you no longer fit its standards.

To finish Day 7, build one simple routine: an action you can repeat weekly without stress. This might be updating your backups, checking your Bitcoin wallet, reviewing your privacy settings, or adding a small piece to your micro-brand. Crisis mode weakens people when they scramble. A steady routine gives you control even in changing conditions. When you know what you will do each week, you stay grounded while the world shifts around you.

By the end of Day 7, your safety net is complete. You have:
- a protected phone and foundation
- strong account and AI boundaries
- a parallel income foothold
- a protected identity

- a backup communication channel
- a self-custodied financial buffer
- an independent speech platform
- a distribution home
- a repeatable routine

This is not a separate life. It is a stronger version of the life you already have. These seven days create a structure that does not collapse when institutions change their rules, tighten their controls, or fail without warning. You have built the beginning of a parallel system, one that supports your voice, your work, your money, and your safety, no matter what happens next. Do not worry if this takes longer than seven days or if you only complete a few pieces at first. The goal is not speed. The goal is growing your confidence and building real security, one steady step at a time.

Conclusion
Your Power in a Closing World

What You Built

Over the course of this book, you built something most people never realize they need until a system fails them: a structure of independence that does not collapse when the world around it becomes unstable. You learned how digital systems tighten slowly, how institutions shift their rules without warning, and how women often feel the pressure first. More importantly, you learned how to respond with clarity instead of confusion. You secured your phone, strengthened your accounts, created boundaries with AI, and built a foundation that protects your everyday life from quiet forms of control.

You also created parallel power, the kind of steady independence that used to come only from owning land or running a business. You built it through small steps, your own identity, your own communication channel, your own financial foothold, your own place to publish and work. None of these structures require you to fight the systems that misunderstand or limit women. They simply allow you to move in and out of them without losing your voice, your access, or your stability. That is real strength. It comes from preparation, not panic.

Most people live inside systems they cannot see. You now see the patterns clearly: friction, verification, shrinking communication, quiet punishment, and forced dependence. Instead of being pushed by these forces, you built a frame that resists them. You created options where most people have none. You built the beginning of a life that does not rely on one platform, one employer, one account, or one fragile identity. This is not a different life. It is a stronger version of your life, shaped by your choices instead of by the decisions of institutions you never agreed to trust.

Where You Go From Here
The path forward is not dramatic. It is steady. Everything you built in this book grows stronger through small, repeatable habits. You do not need to become an expert in security, finance, or technology. You only need to keep the structures you created in motion. Review your privacy settings once a month. Add something small to your micro-brand when you have time. Keep a little independent money on hand. Use your parallel identity when you want space to think and speak clearly. These are not burdens. They are quiet forms of freedom.

As the world shifts, pay attention to early signs of tightening, extra verification, new demands for personal information, missing support channels, or sudden account restrictions. Do not respond by giving away more of yourself. Respond by stepping back, slowing down, and using the alternate tools you already built. This awareness is not fear. It is recognition. When you understand how systems behave, you stop taking their actions personally. You move more confidently, and you protect more of what matters.

Most of all, continue building at your own pace. Your safety net will expand naturally as you use it: clearer communication, stronger privacy, steadier income, and a voice that cannot be quietly erased. Women have always adapted faster than the systems around them. What you created in these pages is part of that long tradition, small, practical steps that give you more space, more protection, and more control over your own life. The world may grow tighter, but you are no longer moving through it unprepared.

About the author

The author lives removed.

Please feel free to burn part or all of this book, safely, as an effigy.

www.ingramcontent.com/pod-product-compliance
Lightning Source LLC
Chambersburg PA
CBHW020946090426
42736CB00010B/1291